THE CAREER RESOURCE LIBRARY

Careers
in
OUTER SPACE:
NEW BUSINESS
OPPORTUNITIES

Edward Willett

The Rosen Publishing Group, Inc.
NEW YORK

Published in 2002 by The Rosen Publishing Group, Inc.
29 East 21st Street, New York, NY 10010

Library of Congress Cataloging-in-Publication Data

Willett, Edward, 1959–
Careers in outer space: new business opportunities /
Edward Willett.
p. cm. — (The career resource library)
Includes bibliographical references and index.
Summary: Presents the skills, education, training, and experience
that can prepare one for a space-related career, with examples
from the lives of astronauts and other space scientists.
ISBN 0-8239-3358-X (library binding)
1. Astronautics—Vocational guidance—Juvenile literature.
[1. Astronautics—Vocational guidance. 2. Astronauts.
3. Occupations. 4. Vocational guidance.] I. Title. II. Series.
TL850 .W55 2002
629.4'023—dc21

2001003157

Manufactured in the United States of America

Contents

Introduction

"Space: the final frontier." The famous opening words of *Star Trek* are more than just a great way to start a television show—they're also an accurate description of what lies beyond Earth's atmosphere.

Space really is the final frontier. Most of Earth has been explored and mapped; people live just about everywhere it's possible for people to live. But beyond Earth there's a whole universe to study and explore—and even colonize.

In fact, we've already begun. Today, thousands of men and women work in fields that are related to the exploration of space. A lucky few are astronauts, the ones who actually travel into space. Many more work on the ground, building the rockets that take the astronauts into space, designing satellites to study space—and Earth from space—or training delicate instruments on the farthest reaches of the universe to learn everything they can about what's out there.

Very few people will ever travel into space. Maybe you'll be one of them. But even if you're not, if you're interested in space, you can find a career that will let you pursue your passion without ever leaving the ground.

A Brief History of Space Exploration

Space exploration didn't begin with the launch of the first man into space. In a sense, it began the first time some unknown ancient human looked up at the stars and wondered what they were.

The ancient Greeks spent a lot of time wondering about the universe and how it was put together. Aristotle, who lived from 384 to 322 BC, developed the common idea that Earth was the center of the universe, and everyone pretty much accepted that for centuries to come—until 1543, in fact. That's when a Polish astronomer, Nicolas Copernicus, published a book suggesting that the Earth (and the other planets) rotate around the Sun, instead of the other way around. More than sixty years later, a German astronomer, Johannes Kepler, published another book demonstrating mathematically that the planets did in fact circle the Sun.

At about the same time, the first telescopes were invented, and in 1632, Galileo Galilei, one of the first astronomers to use the telescope in his work, published his evidence for Copernicus's idea. The Roman Catholic Church was not amused because Galileo's work seemed to contradict the Bible, and so the Church forced Galileo to withdraw his statements.

But Galileo was right, of course, and in 1687, Isaac Newton published a very famous book called *Principia*, which explained how gravity holds the universe together.

With that basic understanding of the universe in place, astronomers began to learn more and more about the stars and planets that filled it. Writers like Jules

Verne, who wrote *From the Earth to the Moon* in 1865, even began to imagine traveling in space.

But the twentieth century was to be the one in which humans were finally able to send machines and eventually themselves into space. Russia's Konstantin Tsolikovsky put forward the idea of using liquid-fueled rockets in space travel in 1903, and in the 1920s, the American Robert Goddard conducted experiments with liquid-fueled rockets that proved Tsolikovsky was on to something.

The Second World War brought great advances in rocket technology, culminating in the V-2, which Nazi Germany used to bombard London. After the war, many of the German scientists ended up in the Soviet Union and the United States, where they continued their research, beginning with captured V-2 rockets.

Then, in 1957, the Soviet Union put the first satellite, *Sputnik 1,* into orbit. *Sputnik 1* was shortly followed by a satellite containing the first animal, a dog named Laika. The United States tried frantically to catch up, but suffered a series of embarrassing launch failures before finally launching its own satellite, *Explorer 1*, in 1958.

The two countries, which were already rivals in international politics, became locked in a race to see who could outdo the other in space. The Soviet Union got off to a head start by launching Yuri Gagarin into orbit in 1961. The United States's first astronaut into space (though he didn't go into orbit) was Alan Shepard, later that same year. Then President John F. Kennedy upped the stakes by proclaiming the goal of putting a man on the moon by the end of the decade.

The United States succeeded in doing just that; Neil Armstrong, commander of *Apollo 11*, became the first man to set foot on the Moon on July 20, 1969. In all, six

Apollo missions landed on the Moon over the next three years. After that, the focus shifted to Earth orbit for manned space flight. But robots, which had already traveled to Mars and other planets in the 1960s, were sent to the farthest reaches of the solar system. The United States's *Viking 1* and *Viking 2* landed on Mars in 1976, and the *Voyager* and *Pioneer* spacecraft sent back fantastic pictures of Jupiter, Saturn, and other distant planets.

The first space shuttle was launched in 1981; five years later, the space shuttle *Challenger* exploded less than two minutes after launch, killing all seven crew members. For a time, there were no shuttle launches. But by the 1990s, shuttle flights once again came to seem almost routine, taking place every few months to launch satellites and conduct experiments—and, in the past few years, to work on construction of the International Space Station, the current focus of manned space flight.

Meanwhile, here on Earth, astronomers continue to make astounding discoveries about the universe, using tools ranging from giant radio telescopes on the ground to the Hubble Space Telescope, an orbiting observatory that provides images far more detailed than any Earth-based telescope can manage.

Robots, too, continue to explore the solar system, even landing and roving around Mars—the planet on which many people would like to see humans land within the next twenty or thirty years.

Just as the American West went from being wild to being civilized and citified, space is beginning to attract not only explorers but people who see it as a place of opportunity. Today, there are telecommunications companies that launch and own their own satellites, companies with plans for hotels in space, even a company that hopes to launch its own roving robot to the Moon.

All of these programs require talented, skilled, and dedicated people. In the pages of this book, you'll learn about many different space-related careers and what you can do now to begin working toward a life in this exciting, fascinating field.

Starting at the Top: Becoming an Astronaut

1

Julie Payette grew up in suburban Montreal watching the 1970s Apollo Moon missions and dreaming of becoming an astronaut. She read every space magazine she could get her hands on. She even pinned a picture of Neil Armstrong to her bedroom wall. "I wanted to be like the astronauts," she said. "I wanted to be an astronaut."

On May 27, 1999, Julie Payette got her wish, when she rode in the space shuttle *Discovery* to become the second Canadian woman in space and the first Canadian to board the International Space Station.

As you'll see as you read this book, there are many other careers in space besides astronaut—but there's also no doubt that the first thing everyone thinks of when they talk about a career in space is being an astronaut.

Becoming an astronaut is not easy; less than 1 percent of the people who apply to the astronaut program actually become astronauts, but it's not impossible.

In fact, becoming an astronaut is easier than it used to be. The original seven American astronauts were all military pilots with engineering training. Today, though, there are many different types of astronauts, with a wide variety of expertise.

Another big difference between today's astronauts and the original astronauts is that the original seven— Gordon Cooper, Virgil "Gus" Grissom, Donald "Deke" Slayton, John Glenn, Scott Carpenter, Walter Schirra, and Alan Shepard—were all white men. Today's U.S. corps of astronauts includes men and women of all races, ages, religions, and ethnic backgrounds.

Types of Astronauts

Within the National Aeronautics and Space Administration (NASA) there are three different types of astronauts, each with different responsibilities and skills: pilot astronauts, mission specialist astronauts, and payload specialist astronauts.

Pilot Astronauts

These experts fly the space shuttle. One pilot astronaut serves as the commander of the shuttle. Like the captain of a ship, the commander is responsible for the vehicle, the crew, the success of the mission, and the safety of everyone on board.

The shuttle commander is assisted by another pilot astronaut who is second in command and primarily responsible for controlling the shuttle. In orbit, when there are no maneuvers to be carried out, the shuttle commander and pilot will often assist with deploying or retrieving satellites or other payload items.

Mission Specialist Astronauts

These experts work with the commander and pilot and are responsible for coordinating onboard operations. For example, they help plan the crew's activities; monitor the use of the shuttle's food, water, and fuel; and conduct experiments. Mission specialists conduct space walks (known as extravehicular activities, or EVAs) and usually operate the shuttle's Canadian-built robot arm. Julie Payette, for instance, was a mission specialist on her 1999 flight.

Payload Specialist Astronauts

These, as their name suggests, are experts mainly interested in the shuttle's payload. They can be specialists trained in the use of scientific equipment, for example, or doctors whose role is to conduct experiments on the other members of the crew to see how they are adapting to space. Unlike the other astronauts, payload specialists are selected by the country or company that provides the payload (although they must be approved by NASA).

How Astronauts Are Chosen

NASA accepts applications from civilians for the positions of pilot astronauts and mission specialist astronauts at any time; military candidates must apply through their service and then be nominated by their service to NASA. Although applications are accepted continuously, candidates are usually selected every two years (although they may be selected at other times if needed).

The basic educational requirement for both types of astronaut is a bachelor's degree or higher from an accredited institution in engineering, biological science, physical science, or mathematics.

Pilot astronaut applicants must also have at least 1,000 hours of pilot-in-command time in jet aircraft, and NASA prefers that they also have experience as test pilots.

Pilot astronaut applicants also have to be able to pass a stringent physical, the requirements of which include vision that is 20/70 or better without glasses or contacts, and correctable to 20/20 with glasses or contacts, in each eye. Their height has to be between 64 and 76 inches.

Mission specialist applicants are required to have three years of related, progressively responsible professional experience as well as a bachelor's degree. Advanced degrees can offset the experience requirement somewhat: A master's degree counts as one year of experience, while a doctorate counts as the full three years of experience.

Mission specialists don't have to pass quite as strict a physical. Their vision has to be only 20/200 or better uncorrected, though it still has to be correctable to 20/20. They can also be a bit shorter than pilot astronauts, with a minimum height of 58.5 inches.

Payload specialists must have education or experience appropriate to the payload or experiment they're in charge of, and they must also pass a physical, the standards for which vary, depending on what they'll be doing in space.

Of course, meeting the basic requirements is only the first small step toward becoming an astronaut. When new candidates are being chosen, qualified applicants are evaluated by panels. Finalists are further screened through a weeklong process of personal interviews, medical examinations, and orientation. The final selection by the Astronaut Selection Board is based very much on personal interviews. Astronauts are expected to work well as members of a team and yet be self-assured and self-reliant individuals. They must also be highly skilled without being so specialized that they can't deal with tasks outside their area of expertise.

The chosen applicants are assigned to the astronaut office at the Johnson Space Center in Houston, Texas, for one to two years of training and evaluation. Civilians who successfully complete that process are expected to remain with NASA for at least five years. Military people are assigned to NASA for a specific tour of duty.

How Astronauts Are Trained

Congratulations! You've made it through the astronaut selection process and now you're at the Johnson Space Center to start your training. What can you expect?

Basically, you're going to school. Astronaut candidates attend a lot of classes. Topics include shuttle systems, basic science and technology, mathematics, geology, meteorology, guidance and navigation, oceanography, orbital dynamics, astronomy, physics, and materials processing. Exactly which topics a particular candidate studies depends on his or her own current knowledge and the focus of upcoming shuttle missions.

It's not all classroom work, of course. Candidates are also trained in land and sea survival, scuba diving, and operating space suits. (In preparation for the sea survival training and for space-walk training, which takes place under water, all candidates have to pass a swimming test during their first month of training. In it, they have to swim three lengths of a twenty-five-meter pool in a flight suit and tennis shoes, and tread water continuously for ten minutes.)

Candidates will be exposed to high and low atmospheric pressures and learn to deal with emergencies related to each, and they will also ride a modified KC-135 aircraft dubbed the "vomit comet" to experience twenty seconds of weightlessness.

Of course, pilot astronaut candidates have to keep up their flying skills. They're expected to fly fifteen hours per month in NASA's fleet of two-seat T-38 jets and also practice landing the space shuttle using a modified Gulfstream II business jet.

Both types of astronaut candidates spend a lot of time working with simulators. These include the single systems trainer (SST), which models one of the shuttle's many systems at a time, and the shuttle mission simulator, which provides training in all aspects of a shuttle mission, from prelaunch to orbit to landing. There are many other partial mock-ups and trainers as well.

Candidates start with generic missions, then move on to specific simulations of their own upcoming mission once they've been assigned to one. Once they have been assigned to a mission, they also work on simulations with the flight controllers assigned to the same mission so that everyone involved learns to work together as a team. A given mission is usually simulated for around 300 hours on the ground before the actual flight.

Once assigned to a flight, pilots spend about 100 hours practicing shuttle landings in the modified corporate jet mentioned earlier; that's the equivalent of about 600 shuttle landing approaches.

All candidates participate in many mission-related technical meetings as their flight dates approach, and in test and checkout activities at the Kennedy Space Center in Florida, the launch site for space shuttles.

The training is so extensive that astronauts often comment that the only thing different between the practice sessions on Earth and the actual flight is the noise and vibration of launch and the experience of weightlessness.

What to Do Now If You Want to Be an Astronaut

Believe it or not, you can start preparing to be an astronaut in elementary school. As Marine Corps Major General Charlie Bolden, himself a former astronaut, puts it, "It is here that the foundations are laid down and then put upon. Start with the basics and get them down first . . . you can't do anything without math and science." He recommends that elementary students read everything they can find about astronauts, space, and their particular field of interest.

But beyond academics, Bolden says that other important skills include knowing how to work well as a member of a team; understanding and appreciating your ethnic, cultural, and American history; and keeping up with current events.

Many astronauts also recommend Scouting as an excellent activity to become involved in because it builds many of the basic skills an astronaut needs.

Once you're in high school, it's important to continue to earn the best grades you can and to make some decisions about what you want to study in college: engineering, biological or physical science, or mathematics. (Remember, you must have a degree in one of those fields to qualify as an astronaut.)

Make sure you pick a field you're interested in anyway. The best way to prepare yourself to be an astronaut is to become very, very good in your chosen field—and you won't be able to do that unless you really enjoy what you're doing.

Consider taking a second language. As Bolden notes, "Space is a multinational and multicultural operation." Many missions involve astronauts from other countries. In particular, astronauts are working closely with Russian cosmonauts in the International Space Station.

Also give careful consideration to which college you should attend. NASA contributes funds to many schools through its Space Grant Consortia. Attending one of those schools and following the curriculum offered for space programs will help ensure that your education meets NASA's standards.

Whatever school you choose, participating in an internship or co-op education program is also an excellent idea; it will give you vital work experience in your field of study. Once you've completed your education and/or gained the experience NASA requires, the time has come to apply. Your goal may seem impossible. But remember Julie Payette.

When Julie dreamed of being an astronaut in Montreal in the 1970s, not only were there no women astronauts, there were no Canadian astronauts. Nevertheless, the career path Payette followed was exactly what she needed to eventually make her dream become a reality. She speaks six languages (including Russian). She's a gifted singer and musician, a triathlete and skier, a scuba diver, and licensed pilot. After high school, Payette attended the International College of the Atlantic in South Wales, UK, and returned home to graduate with a bachelor of engineering degree from McGill University in 1986 and a master of applied sciences degree from the University of Toronto in 1990. In 1992, while working for Bell-Northern Research in Montreal, she saw a newspaper ad calling for new recruits to the Canadian Space Agency. There were 5,300 applicants.

Asked during one interview how she could demonstrate her ability to work in a team, Payette said she sang in several choirs. "You have to listen to what the others do," she said. "That is really exquisite teamwork."

In June 1992, Payette was officially named one of four new Canadian astronauts.

No, it's not easy to become an astronaut. But if you prepare yourself well and work hard to achieve your goal, maybe you, too, will be paying a visit to the International Space Station in the next few years.

The astronaut application form can be obtained from the Astronaut Selection Office, Mail Code AHX, Johnson Space Center, Houston, TX 77058-3696.

Learning More About and from Space: Physical Scientists

You can be fascinated by space and want to learn every-thing that you can about it—or from it—without ever actually traveling into it. Instead of becoming an astro-naut, you can become a physical scientist: someone who studies how the universe is put together.

There are many different types of physical scientists, more than can be mentioned in this book. Not all of them are involved in studying space, of course; some of them are entirely focused on topics here on Earth. But just about every branch of physical science includes researchers who are either studying some space-related topic or using space technology in their research.

In this chapter, we'll look at a half dozen different types of physical scientists whose work involves space.

Astronomers

In any list of space scientists, astronomers have to come first. Astronomers are scientists who study the planets, stars, and galaxies that lie in outer space. Before the telescope was invented, astronomers had to base their understanding of what lay beyond Earth on what they could see with their naked eyes; today, astronomers use many different types of instruments for observing space. In addition to regular earthbound optical telescopes, they use space-based telescopes, like the Hubble Space Telescope. They also use other instruments, both on the ground and in orbit, to detect other forms of electromagnetic radiation from space. These include radio telescopes—essentially giant radio receivers—gamma-ray telescopes, infrared telescopes, and more. Some even use enormous underground vats filled with cleaning fluid to try to detect elusive particles called neutrinos, which fly through the earth without stopping but occasionally collide with an atom of the cleaning fluid and set off a tiny spark of light that astronomers can detect and measure.

Astronomers have discovered and continue to discover many strange and exotic things in outer space, such as black holes, exploding stars, colliding galaxies, and planets orbiting other suns. Some astronomers are even involved in listening for radio signals from possible alien civilizations in other parts of the galaxy.

In a sense, astronomers journey into deep space all the time, without ever leaving the ground.

Dean C. Hines, an astronomer at the University of Arizona, says he has wanted to do astronomy since he was a little kid. "When I was a child, my parents sat me in front of the TV to watch the space flights," he says. "I didn't miss a single launch until the eighth shuttle liftoff! My dad bought me a telescope when I was five, and we often used it to look at the planets, or we'd lie on the hood of the car and just look up." He adds, "There are some who stumble into astronomy late in the game, but the majority of us have always had our eyes on the stars."

Geologists

At first glance, you might think geologists wouldn't care about space at all or have any use for it in their research. Geologists study the rocks that make up Earth and try to figure out how it formed and how it is changing.

Before space flight became possible, geologists had to rely on what they could observe from the ground, or at best from an airplane a few thousand feet above the ground. But today, many geologists rely heavily on images taken of Earth from space. Analyzing these images tells geologists much more about the processes that are shaping the planet, from volcanoes to earthquakes to the movements of glaciers and even continents. In addition, some geologists are involved in the astronomers' work of studying other planets. Learning more about how the Moon or Mars or Venus or Mercury came to be the way they are also tells us more about Earth's evolution.

Jeff Plescia is a geologist who works as a research scientist at NASA's Jet Propulsion Laboratory. His focus is on both Martian geology and meteorite craters found on Earth. He has also been very involved in planning for future missions to Mars—both the robot ones that are under way now, and possible future manned missions.

"I remember sitting in front of the television watching with great interest and excitement the *Apollo 11* landing on the Moon and following the entire space program—the manned and robotic," Plescia says. "I've long had an interest in space and that's what brought me into planetary geology."

Some geologists have even become astronauts, such as Dr. Harrison "Jack" Schmitt, one of the *Apollo 17* astronauts who spent more than three days in December 1972 roaming around the surface of the Moon on the dune buggy–like Lunar Rover.

Meteorologists

Meteorologists are people who study Earth's weather in order to forecast changes. They're the ones who tell you it's going to rain on Saturday or to cover your tomatoes because of the risk of frost.

They're also the ones who warn people of tornadoes and hurricanes and other extremely dangerous weather events. These kinds of terrible storms used to cause even more damage and loss of life than they do now. Today, satellite images and sensor readings provide meteorologists with the data they need to make much more accurate predictions than they used to be able to. Satellites provide information about clouds, radiation, winds, dust, and precipitation.

One branch of meteorology, climatology, is becoming particularly important because of the threat of climate changes caused by the warming of the planet due to the increasing levels of carbon dioxide and other gases in the atmosphere. Climatologists, too, depend heavily on information provided by satellites as they try to determine just how global warming may affect the planet.

All meteorologists, even if they're working as TV weather forecasters, are involved with space technology because they rely on satellite photographs and other information from space. But some meteorologists are even more directly involved; they work for NASA, either as research scientists or predicting weather that might affect launches.

NASA meteorologist Marshall Shepherd thinks it's a great career. As he puts it, "I get to come in to work and use all these fancy toys that NASA has: aircraft, space satellites, and so on. They pay me to do something that I would like to do as a hobby anyhow."

Oceanographers

Oceanographers are scientists who study the oceans and the creatures that live in them. Like geologists, they might at first seem to have no reason to be interested in space; like geologists, however, they rely on images and other readings from space for vital information about what the oceans are doing.

Satellites can provide information on the heights of waves all over the ocean, water temperatures, pollution, and more.

One current NASA project, for instance, is the Sea-viewing Wide Field-of-view Sensor—SeaWiFS, for short. Carried aboard the *SeaStar* satellite that was launched in 1997, SeaWiFS looks at the color of the ocean, and by doing so, reveals what's in it—whether that's plankton or pollution.

Among other things, SeaWiFS data is helping us better understand El Niño, a warming of the Pacific Ocean that has a major influence on North American weather. More than 800 scientists, many of them oceanographers, representing thirty-five countries, have registered to use the data.

"With this record we have more biological data today than has been collected by all previous field surveys and ship cruises," says oceanographer Gene Carl Feldman, SeaWiFS project manager at Goddard Space Flight Center. "It would take a ship steaming at six knots over 4,000 years to provide the same coverage as a single global SeaWiFS image."

Data like that, available only from satellites, is why oceanography can also be a career in space research.

Physicists

Physicists, as their name implies, study the physical properties of the world around us. they use data gathered from space in their research, and conduct research that leads to better space vehicles.

Michael D. Hogue is a physicist at the John F. Kennedy Space Center who works to develop automated systems that make use of artificial intelligence—robots, in other words. He became interested in science as a boy, but not right away. "I was pretty ambivalent about science until I was twelve years old and saw my first episode of *Star Trek,*" he says. "I was hooked. I could not get enough books or magazines about spacecraft and space flight. My mother was not surprised when I told her that I wanted to major in physics in college."

There are many different types of physicists. High-energy physics, particle physics, fluid physics, and optics are just some of the branches of physics vital to the space program. Among other things, physicists design experiments to be carried out by space probes and on the space shuttle. They study topics ranging from how the Sun works to the atmospheres and magnetic fields of planets to how materials behave in weightless conditions.

Chemists

It's pretty easy to see one way that chemists play a role in the space program. Every time a rocket launches, the fiery exhaust is the result of chemistry, which is the study of the different elements that make up our world and how they interact with each other.

But chemists are also involved in analyzing data from space probes sent to asteroids, comets, and planets, helping to determine what elements those various bodies contain. Knowing what raw resources are available elsewhere in the solar system is crucial if humans are to visit and eventually live on other planets.

Chemists also play a role in analyzing information gathered in space about Earth, such as the state of the layer of ozone, high in the atmosphere, that blocks dangerous ultraviolet radiation from the surface, and the results of various kinds of air pollution.

Other chemists work in the field of materials science, trying to find better materials out of which to build space technology. Dionne Broxton Jackson, who works at the Kennedy Space Center, tests and identifies metal alloys (particularly important in Florida since the seaside location means a constant threat of corrosion). She knew she wanted to be a chemist when she graduated from high school, and she made that decision because of her participation in NASA's Summer High School Apprentice Research Program. She didn't know when she finished that program that she'd end up working at NASA again.

What You Can Do Now If You Want to Be a Physical Scientist

Learn everything you can about the field you're interested in right now. Obviously, you should aim to get the best grades you can, not only in science classes but in all your classes; you need good grades to get into the universities with the best science faculties.

Even if you're not yet far enough along in school to take classes that are specific to the field you're interested in—chemistry or physics, for instance—you can learn a lot about that field outside of class. Read books on the field of science that interest you. Surf the Web; you'll find lots of information about all kinds of science, including the home pages of working scientists, where you can discover exactly what real chemists, physicists, astronomers, and others have done and are doing in their careers. Look for science-related extracurricular activities in which you can participate. Maybe your school has a science club, or maybe a local observatory offers free tours of the night sky on a regular basis. These are quality programs in which you can learn about and experience science projects.

Dr. Sten Odenwald, an astronomer with Raytheon Information Technology and Scientific Services in Washington, D.C., notes, "By the time I had finished ninth grade, in my after-school hours I had built from scratch two radios, a transistorized telescope 'clock drive,' and a 4 1/2-inch reflecting telescope; read every book in both the local public library and the city of Oakland main library on astronomy; took up 'astrophotography' and taught myself how to develop black-and-white film and make prints; and created a simple photo atlas of the constellations."

Dr. Odenwald's experience proves one thing: It's never too early to develop a passion for the field of science you want to pursue as an adult.

Once you're in high school, of course, you should continue to concentrate on taking science courses. Some high schools also offer advanced science courses, which can be of great benefit when the time comes to go on to university.

Long before you graduate from high school, you should already be exploring which universities offer bachelor's degrees in your chosen field. Make full use of any help offered by your school guidance or career counselors.

Once you have a bachelor's degree, work toward a master's degree in some space-related area of specialty. Check into the possibility of internships in your chosen field. Eventually, if you're interested in pursuing your own independent research, you'll need a doctorate (Ph.D.) degree as well.

It's not always easy to find employment in your chosen field, even with an advanced degree. And even if you can find work as, say, a geologist, you may not be able to find work that also allows you to be involved in space research. But if you keep your eyes on your goal and take advantage of opportunities as they arise, someday you may have the best of both worlds: your feet planted firmly in a solid career on Earth, but your head just as firmly in the stars.

Studying Humans in Space: Life Scientists

3

When Dr. Roberta Bondar was a child in Sault Ste. Marie, Ontario, she used to gaze up at the stars and dream of traveling in space. But Dr. Bondar was Canadian, and for a long time there were no Canadian astronauts.

Instead, Dr. Bondar became a scientist. She studied zoology and biology in college, working her way up to a Ph.D. Then she went to medical school and became a medical doctor, specializing in neurology, the study of the brain, and specifically in neuro-ophthalmology. Along the way, she also became a pilot; in fact, she knew how to fly a plane before she had a license to drive a car.

In December 1983, six Canadian astronauts were chosen from thousands of applicants. Dr. Bondar was one of them. In 1990, NASA chose her to become the payload specialist in charge of the experiments for the first International Microgravity Laboratory Mission—a mission designed to study the effects of weightlessness on the human body.

Dr. Bondar flew on the space shuttle *Discovery* from January 22 to 30, 1992, becoming both the first Canadian woman and the first neurologist in space.

Dr. Bondar is just one example of how life scientists—biologists, medical doctors, physiologists, nutritionists, and psychologists, among others—play an important role in the space program.

Some life scientists, like some physical scientists, are involved with space exploration through the use of data from space in their research—tracking algae blooms in the oceans through satellite imaging, for instance. But the main focus of life scientists within the space program is the effect of space flight on living organisms.

This has been an interest of space scientists since the beginning of the space program. The first living thing in orbit, after all, was a dog, Laika, launched by the Soviet Union in *Sputnik II* on November 3, 1957, and closely monitored to see how space flight affected her.

In the United States, chimpanzees rode into space in Mercury space capsules before astronauts did, for the same reason. And once astronauts did start traveling into space, they were wired with all kinds of sensors so scientists on Earth could keep a close watch on how they fared.

Eventually, doctors themselves went into orbit to conduct experiments on astronauts. The first doctor in space was Boris Borisovich Yegorov, one of three cosmonauts aboard the *Voskhod 1* flight, which launched October 12, 1964, and marked the first time more than one person had been launched into orbit. Dr. Yegorov took blood samples from his crewmates while in space.

Many more doctors and other life scientists have traveled into space since then; many, many more continue to work on the ground to understand the effects of space flight on living organisms, especially humans. We need to understand those effects thoroughly if we're ever going to undertake long flights to the other planets of the solar system—or beyond.

How Space Flight Affects Humans

In the movies, spaceships have artificial gravity because it's a heck of a lot easier to film that way. Real-life astronauts aren't so lucky; they're weightless all the time they're in space, which in the International Space Station can be months.

Sure, weightlessness can be a lot of fun—flying around, leaving objects hanging in mid-air—but inside the astronauts' bodies, things are happening that aren't much fun at all.

Consider Dr. David Wolf, who at age forty-one spent four months aboard the Russian space station *Mir*. Before he left Earth, he could run a six-minute mile. When he returned, he could hardly walk.

That's because nearly every system in the body is tied to and affected by the force of gravity. When that force is removed, those systems change in ways that could spell serious problems for long-duration space flights.

The most worrisome problem is one we mostly associate with the elderly here on Earth: osteoporosis, a weakening of the bones. When gravity is removed, the bones immediately began to lose calcium, which is absorbed into the body.

The minerals lost from the legbones and hipbones aren't excreted. Instead, they migrate, primarily to the head. The body, in other words, is making better use of its resources. Since the legs aren't being used, the body quits worrying about them and uses the freed-up resources to better protect that most vital of organs: the brain.

On Earth, we build bones by running, jumping, and weight lifting—anything that loads them with more than the body's normal weight. One of the goals of life scientists studying the human adaptation to space flight is to find a way to mimic the effects of that exercise in space. Exercise bikes and treadmills to which astronauts are held by bungee cords haven't proven very effective. If an effective substitute for Earth-type exercise can't be found, astronauts on long missions in space may have to begin taking drugs that block the reabsorbing of calcium from the bones, the same drugs taken by many elderly people on Earth.

Bones aren't all that weakens in the body during the time spent in space. Muscles lose protein, size, and strength. Exercise programs developed so far have been only partially successful in stopping this process, which is one reason returning astronauts feel so weak.

The cardiovascular system is affected, too. On Earth, gravity pulls blood to the lower body, away from the head. Special nerves detect the difference in pressure and cause the body to redirect blood flow, ensuring the brain gets the oxygen and sugar it needs to operate. In space, the nerves don't sense any pressure difference. The brain interprets that as meaning the body is bloated and tells the body to increase urination.

Over time, the nerves forget how to respond to gravity at all. As a result, when the astronaut returns to Earth and tries to stand, no extra blood is sent to the brain, and the astronaut feels faint.

Astronauts also grow a couple of inches taller in space because gravity is no longer squashing their spines. Many astronauts suffer excruciating lower-back pain as a result of stretched ligaments.

Astronauts may also suffer from anemia; the number of oxygen-carrying red blood cells may drop as much as a third. As well, many astronauts suffer from space sickness, probably because what their eyes tell them doesn't match the messages from their confused inner ears. Back on Earth, spatial orientation can suffer as the body readapts to gravity, which can cause astronauts to stagger when they try to walk.

The gut becomes less effective at absorbing food. The immune system suffers. Sleep is disrupted. Biological rhythms change—body clocks get out of synch with those of ground control, which affects job performance. Even temperature regulation can get out of whack.

There are also psychological effects to life in space, where you have a small group of men and women, sometimes of different nationalities, living and working together in very close quarters under often stressful and sometimes uncomfortable conditions.

There is still a lot to understand about how humans react to life in space and how to counteract some of the negative effects. We also need to understand more about how other forms of life react to the space environment; any long flights to other planets will probably require the crew to grow some of their own food in the form of plants.

Learning more about how living things are affected by space will also teach us new things about how living things adapt to new conditions on Earth. We could learn more about the human immune system, for instance, which could lead to improved treatment of diseases.

All of this means there will continue to be many exciting opportunities for life scientists within the broader field of space research.

Of course, not all life scientists involved in the space program are working on space adaptation syndrome. There are also many space-related career opportunities for life scientists on the ground.

Physicians

Many physicians work for NASA, some researching the effects of space flight on humans, others looking after the health of astronauts and other NASA employees on the ground. One of the latter is Dr. George Martin, who works at the Kennedy Space Center.

"My primary job is to maintain the health of the astronauts and the workers who support the astronauts," Dr. Martin explains. "My office is also responsible for making sure that none of the work done out here at KSC damages the animals and plants and the rest of the natural environment."

His work is varied; besides training other medical staff and dealing with medical problems, he sometimes participates in astronaut training and sometimes goes aboard the shuttle when it has just landed for the initial checkup of the flight crew. Sometimes, he even goes out to the shuttle just before launch to test the quality of the drinking water.

Says Dr. Martin, "I think that I have one of the best jobs any doctor could ever have."

Psychiatrists

As mentioned earlier, there are psychological as well as physiological problems faced by space crews, who, especially on board the International Space Station or future long-distance flights to Mars or other planets, spend months in very tight quarters. That means there are space-related career opportunities for psychiatrists and psychologists, too.

Dr. Nick A. Kanas of the University of California, San Francisco, is a psychiatrist who has spent part of his career studying astronaut and cosmonaut crews. "It is hard to live with two or three people in space for months," Dr. Kanas says. "If you have a fight on Earth, you can go out to a movie or something and it is all better. But in space, you can't go anywhere. You are there." Over time, crew compatibility breaks down, some members gang up on others, leaders may begin to lose their authority, and everyone develops a tendency to blame mission control for problems.

If this is a problem in space stations over missions lasting just a few months, it will be even more of a problem for missions to other planets lasting one or two years. That means there's lots of research to be done to better understand how crews interact in space, and how crew members can be screened on the ground to ensure their compatibility once they leave Earth.

Nutritionists

Nutritionists study the science of food, and many nutritionists work at NASA, trying to determine exactly what foods and nutrients astronauts need to keep them in peak condition while in space.

Janis Davis-Street is a NASA nutritionist working at the Kennedy Space Center on exactly that problem. She grew up in Guyana, went to college in Canada, and immigrated to the United States in 1987. Though she knew in high school she wanted to be a doctor and has always had an interest in science, she says, "I would never have dreamt that someday I would be involved in the United States space program." As Davis-Street says, "anything is possible in life with hard work, determination, and emotional support."

Biologists

Biologists are scientists who study living things: bacteria, plants, animals. Many biologists are involved in the study of the effects of space on living organisms, as mentioned earlier. For instance, spiders have been flown into orbit to see how they spin their webs in the absence of gravity.

Other biologists conduct research on life on Earth using satellite information to further their studies. This includes information from the SeaWiFS satellite described in the last chapter that tells biologists something about the amount of plankton in the oceans.

Others have more unusual careers. Take John Rummel, for instance; he's the planetary protection officer. It's his job to make sure that spacecraft don't return from outer space or other planets carrying potentially dangerous alien life-forms. He's kind of like one of the characters from the movie *Men in Black,* only without the black suit and shades and definitely without the little pocket device that makes everyone forget what they've just seen. Instead, John relies on careful sterilization of space probes going to other planets (so our germs don't contaminate them) and quarantine of samples returning to Earth (like the samples from Mars scheduled to be collected by a mission planned for 2005).

The position of planetary protection officer is part of Rummel's larger role as manager of NASA's Exobiology

Program, which is charged with searching for life elsewhere in the universe—something he's convinced exists.

"I think that there's life elsewhere in the universe, and no reason to think otherwise," Rummel says. "Certainly, everything that appears to have lead to life on Earth is taking place elsewhere in the galaxy—this galaxy, not just elsewhere in the universe. There are 400 billion stars in this galaxy alone. To say that we're the only place where any of this stuff could possibly have happened is essentially absurd. We're going to have all sorts of opportunities for life elsewhere."

What to Do Now If You Want to Be a Life Scientist

Becoming a life scientist is no different from becoming a physical scientist. As discussed in the last chapter, you should read and learn everything you can about your chosen field. Concentrate on getting good grades, especially in science classes (and especially, obviously, in biology), then go on to a good university to obtain a bachelor's degree and eventually a postgraduate degree in a space-related specialty.

With hard work, perseverance, and luck, in a few years you may be one of the doctors looking after the crew of the International Space Station, or even the crew of the first manned flight to Mars.

Maybe, like Dr. Bondar, you'll even make the trip into space yourself. Or maybe, like John Rummel, you'll be the planetary protection officer, charged with protecting Earth from infection by alien organisms.

Mathematics Makes the Spacecraft Go Round

If you're studying mathematics in school right now, or if you've ever taken a math class, you've probably heard someone say, "What good is all this stuff in the real world?" Maybe you've even said it yourself.

Maybe Katherine Goble Johnson asked herself that same question when she was attending school in rural West Virginia in the 1920s, in a town where black children like her were not allowed to attend school with white children.

Fortunately, Johnson's family valued education enough to move 120 miles to Institute, West Virginia. There she was able to attend high school, and in 1937 she graduated from West Virginia State College with a degree in mathematics.

Katherine Johnson taught school for many years, then in the 1950s joined the National Advisory Committee for Aeronautics (NACA) as a pool mathematician. Within a few years, NACA became NASA, and Johnson became part of the space flight research group at NASA's Langley Research Center in Virginia.

When America sent its first astronaut, Alan B. Shepard Jr., into space in 1961, Johnson was the person who had calculated his trajectory. And when the Apollo astronauts reached the moon in 1969, they did so thanks to Johnson's calculations. That's one thing that math is good for in the "real world"!

A space probe that's launched one year and arrives at Mars or Jupiter months or years later doesn't get there through luck; it arrives because of the work of mathematicians.

Why is math so important to space flight? Because math is essentially a language that can be used to describe the world around us. In space science, mathematics is used to describe the whole universe. By writing the laws that govern the orbiting of the planets and the trajectories of spacecraft in mathematical language, scientists can tell where spacecraft will be at any given time just by plugging in details of the specific planets and spacecraft involved.

As you might expect, it takes some pretty complex mathematics to describe the universe. Mathematicians involved in space exploration use all the many mathematical tools that have been invented over the centuries, including algebra, geometry, number theory, calculus, and logic.

And while the space-faring heroes of science fiction fifty years ago made do with slide rules, the real-life mathematicians of today's real-life space program use computers, which means among the mathematics-based careers available in space exploration are such computer-centered jobs as computer programmer and systems analyst.

Even people whose job title is just "mathematician" or "statistician" will find that today they use computers all the time, and may often have to program the computers themselves to analyze their problems properly.

Let's take a closer look at some of these mathematics-based careers and how people in these careers are active in the space program.

Computer Scientists

The term "computer scientist" actually covers a wide range of computer professionals who design computers and their software, and develop ways to apply computers to new uses. Within the space program, for example, there is a constant need for specially designed computers to carry out specific tasks and withstand the rigors of space flight. Computer scientists figure out how to create these special computers. They also try to figure out ways to use computers to simplify other space flight-related tasks.

Computer scientists are also involved in developing computers to analyze the data received from space so that some of the other types of scientists we've already talked about can make the best use of the information that satellites and other space probes send back to Earth.

Carol B. Davies is a computer scientist at NASA's Ames Research Center, where, she says, "I develop mathematical methods and write computer codes to help in the design of space vehicle configurations. I use a lot of math, especially 3-D geometry and vector analysis." Fortunately, she adds, "I was always good at math."

Davies says the best part of her job is "being able to work alone on solving problems, but having a vast number of highly qualified people to provide help whenever needed."

Systems Analysts

A systems analyst helps solve computer problems and helps organizations use their existing computers better. Systems analysts might help NASA develop better ways of linking their many different computer systems together, for instance, or improve the efficiency of communication among tracking stations around the world.

Sometimes systems analysts design whole new systems, or new software for existing systems, to improve efficiency and capabilities. In general, while a computer scientist is more likely to be focused on a single computer, systems analysts are more interested in the ways in which many computers work together.

Statisticians

Statistics is a special branch of mathematics that focuses on the collection, analysis, and presentation of numerical data. Statisticians are therefore often involved with other types of scientists, helping them to design experiments that will generate the kind of data they need to answer a particular question, then helping them interpret the results of those experiments.

Within the field of space exploration, statisticians are valuable not only because they can help other scientists interpret the data they receive, but also because their skills can be used to help determine, for instance, how often certain systems can run safely without being replaced. If told how often a representative sample of, say, rocket engines lasted until they failed, a statistician can predict how long any such rocket engine is likely to last without failing.

What to Do If You Want to Be a Mathematician

To begin with, you should be good at math. More than that, you should really enjoy it. When I was in school, I knew a student who used to create and solve complex algebraic equations just for fun. If you're that kind of student, mathematics may be the ideal career field for you.

Again, read everything you can about mathematics and the history of math. Mathematics is one of the oldest sciences, dating back to the ancient Greeks and Egyptians and even earlier, and it has a fascinating history; you won't be bored.

In high school, take all the mathematics courses that are offered: algebra, trigonometry, calculus, geometry, statistics, whatever is available.

Learn about computers—not just how to play games on them, or surf the World Wide Web, but how to program them. Good computer skills are essential for the modern mathematician.

Be prepared to stay in school for a long time, especially if your focus is on mathematics, and not on one of the related computer fields like computer scientist or systems analyst.

Most people who are actually hired with the job title "mathematician" have a Ph.D. in mathematics, while most statisticians have at least a master's degree.

To work in the space industry, it also helps to have related experience or knowledge in some of the physical sciences, such as those described in chapter 2.

Finally, it's important to be able to think logically and have good communication skills. Even before you get to college you can work on those abilities. Debate, drama, and creative writing classes or clubs are a good way to improve communication skills. Debate clubs are also a good place to learn logical thinking.

Ultimately, your degree, experience, and enthusiasm for mathematics-related careers may earn you a place in the space industry, with NASA or with one of the many commercial and research organizations designing space systems or collecting and interpreting data from space.

Then, like Katherine Goble Johnson, you, too, will have found an excellent way to make use of all that "math stuff" in the "real world"—and out of it.

Making Things Work: Engineers

When Donna Shirley walked into her adviser's office on her first day at the University of Oklahoma and told him she wanted to major in aeronautical engineering, she was told her dreams could never come true. "Girls can't be engineers," she was told.

But on July 4, 1997, when the *Mars Pathfinder* spacecraft became the first robotic probe to land on Mars since the *Viking* landers of the 1970s, Donna Shirley, coordinator of NASA's entire Mars exploration program, was in charge of the project.

Shirley, who grew up in the small Oklahoma town of Wynnewood, was inspired by science fiction books like Arthur C. Clarke's *The Sands of Mars.*

"I just got completely fascinated with the idea of living on Mars," she says, and so she set out to become an engineer.

Determined to prove her adviser wrong, she almost proved him right: She took so many engineering classes she got swamped in the schoolwork and switched her major to journalism. But after spending a short time as a technical writer, she returned to engineering, getting her master's degree.

Her first engineering job was designing the aerodynamic shape of a proposed Martian vehicle for a private company. But she realized that any realistic Mars mission would go through NASA, and so she moved to Pasadena, California, to be near the Jet Propulsion Lab.

In the 1980s, she designed Martian roving vehicles at JPL and headed up a NASA panel that set out tough recommendations about how NASA should approach future Martian missions. Those recommendations were embraced by Dan Goldin when he took over as NASA's chief administrator, and Shirley's dream of exploring Mars came true with the successful landing of the *Mars Pathfinder* probe and its roving counterpart, *Sojourner.*

What Is an Engineer?

An engineer is a person who has the skills to put scientific knowledge to practical use. Or, to put it even more succinctly, engineers are the people who make things work. There are many different branches of engineering, and many different types of engineers play important roles in the space program.

Aeronautical Engineers

Like Donna Shirley, aeronautical engineers design aircraft and spacecraft, rockets, and satellites, and supervise their manufacturing, testing, modification, maintenance, and repair. They also design engines and propulsion systems for aircraft and spacecraft. They have to have a firm understanding of both aerodynamics and mechanics.

Avionics/Instrumentation Engineers

Avionics/instrumentation engineers specialize in designing, developing, manufacturing, testing, and installing the electronics, instruments, sensors, computers, and flight controls used in aircraft and spacecraft.

Spacecraft Engineers

Spacecraft engineers design the various systems that go into the spacecraft that the aeronautical engineers design, such as telemetry and communications, instrumentation, power supplies, and life support systems.

Materials Engineers

Materials engineers are involved in the development and application of the many special materials used in spacecraft to help them withstand the enormous forces and stresses involved in launch, orbiting, and landing. One example is the ceramic tile that covers the bottom of the space shuttle to protect it from the heat of reentry, generated by friction with the atmosphere. (Actually, ceramics—one of humanity's most ancient crafts—is used extensively in spacecraft. Glass fibers that carry information in the form of pulses of light are a form of ceramics, too.) Other special materials protect spacecraft from the heat, cold, and radiation they're exposed to in space.

Computer Engineers

Computer engineers design the computer systems, including the hardware, software, and networking equipment, that control spacecraft and gather data from them. (NASA's computer engineers got a workout during the STS-100 flight in April, 2001, when the International Space Station's main computer and its two backups all crashed at once, delaying the completion of the mission.) Some computer engineers are primarily computer programmers.

Robotics Engineers

Robotics engineers design and build robots to do things that would be dangerous or impossible for humans. Robots can work in the vacuum of space, work in a hazardous environment on Earth, or even be landed on another planet. (As a rule, real robots don't look like the mechanical men of science fiction movies. All space machines are robotic. So is the mechanical arm the shuttle uses to move objects in and out of its cargo bay while in space.)

Electrical Engineers

Electrical engineers plan and develop electronic controls and communications equipment for satellites and manned spacecraft.

Specialized Antenna Engineers

Specialized antenna engineers concentrate on the design of the antennas that allow spacecraft to communicate with Earth. These people are both electronics and communications experts.

There's room for almost any type of engineer in the space program, primarily because almost everything used in space is designed and built—engineered, in other words—from scratch. You can't just go down to Wal-Mart and buy a space suit in the clothing department.

Instead, you turn to someone like Phil West, who's a space suit engineer. Technically, West is a mechanical engineer; that's what his degree is in. But over the years, his career has focused almost exclusively on space suits and the tools astronauts use in orbit for construction jobs like the International Space Station.

He has conducted tests and developed space suits, and worn them himself in vacuum chambers and in the "vomit comet," the airplane NASA uses to create weightlessness.

So far, you can't get a degree in space suit engineering, but designing space suits is just one of the exciting career possibilities available for engineers who choose to work in the space industry.

What to Do If You Want to Become an Engineer

First of all, as Donna Shirley proved to her adviser, you don't have to be a boy to become an engineer. Although you'll still find more men than women in college engineering courses, the number of women pursuing engineering degrees is increasing all the time.

What you do have to have is a strong interest and high skill level in mathematics and science. All engineering is built on those two fields of study. You need to learn how to study well and remember what you read. You should be curious about how things work. And because engineering, as a profession, considers its highest priority to be the protection of the public, you should also think about the social implications and consequences of everything you do.

Communication is also an important part of engineering because engineers, especially in the space program, don't work alone. Engineering is a "team sport," as Donna Shirley puts it. "It's very exciting to be part of a team building things," she adds.

Learning to work as part of a team means learning to seek out other people's opinions and knowledge when you have a problem to solve, and also learning to negotiate and find compromises. So anything you can do now to improve your ability to communicate—writing for the school paper, for instance, or studying drama—will help you if you decide to pursue a career in communication. (Shirley, remember, actually studied journalism and worked as a technical writer before coming back to engineering.)

Engineering is also a very creative enterprise. Its goal, after all, is to solve problems and to figure out the best way to accomplish some task or meet some need. Whenever you're given a problem to solve in school, see how many solutions you can come up with.

Reading a lot can help improve your creative ability because it gives you the opportunity to see how other people solve problems. Maybe they're make-believe problems like the ones in science fiction and fantasy books, or maybe they're real problems like the ones in history and science books. Either way, reading is a workout for your imagination and will make you a better engineer in the future.

Many engineers are also creative in other ways: they may be artists or musicians or actors. Math and science are very important, but don't focus on them to the exclusion of everything else. The time you spend painting or songwriting may someday help you see a new solution to a difficult problem, and may one day save a billion-dollar space mission.

In high school, of course, you should take all of the math and science courses you can, and begin looking for a college with a good engineering program.

Once you're in college, you'll probably start by taking the same classes that all the other engineering students are taking, then decide at the end of that year what kind of engineer you want to be. Over the ensuing four years, you'll take classes more specific to your specialty (and, as you've seen, almost all types of engineers are needed for space-related activities), as well as various elective courses. Your goal is a bachelor's degree, after which you may be able to go to work immediately in engineering or may choose to pursue an advanced degree, which doesn't necessarily have to be in the specialty you focused on for your bachelor's. Keep in mind that your goal is to work on space-related projects, and choose your employer or further training accordingly.

Scientists provide the basic knowledge we need to travel into space and decide what we should do while we're there; astronauts do the traveling; mathematicians draw the road map; but it's the engineers that actually make space exploration happen.

If literally building our presence in space appeals to you—if, for example, like Donna Shirley, you're obsessed with the idea of living on Mars—then engineering may be the career for you.

Keeping Things Running: Technicians

Glen Davis has never traveled into space, but he has probably been inside the space shuttle more often than any astronaut. That's because, as a mechanical technician who looks after shuttle systems at the Kennedy Space Center, Davis is the last person to leave the shuttle before fueling and launch. When a shuttle lands, he's the first person inside to take over responsibility for the shuttle from the astronauts.

Davis works for United Space Alliance, the private contractor that maintains the shuttle fleet for NASA. He's responsible for the performance, checkout, verification, modification, and maintenance of major shuttle mechanical, hydraulic, propulsion, and pneumatic systems. Before launch, he serves as the eyes, ears, and hands of the engineers and others who, sitting at consoles half a mile away in the Launch Control Center, test the orbiter for launch readiness and make sure that all the switches and circuit breakers inside the crew compartment are set the way they're supposed to be.

Davis also works on the outside of the orbiter. Among other things, he helps maintain the nose landing gear and remove and replace the tires. He even does windows, cleaning and polishing them between flights, and cleans the onboard toilet and vacuums the shuttle's floors.

Davis is only one of thousands of technicians and technologists (the two words mean essentially the same thing) who work in all aspects of the space industry. They're people with highly developed technical skills who maintain and often operate the machinery of space flight.

They're people like Ron Woods, who began working in the space program in 1967 as a space suit technician, taking care of all the space suits worn by astronauts on their various missions and helping the astronauts suit up before flights.

They're the "pad rats," who are people who work at the launch pads. Like Bill Williams, lead mechanical technician at the NASA Kennedy Space Center, they prepare the shuttle for flight—filling storage tanks on the orbiter with liquid hydrogen and liquid oxygen to stoke the fuel cells that provide it with power, for instance. The pad rats also look after all the systems on the 295-foot-tall launch tower—no matter what the weather!

They're people like Barry Slack, a biomedical technician at the NASA Kennedy Space Center, who, among other things, equips Black Hawk rescue helicopters with all the high-tech medical equipment necessary to turn them into flying ambulances that can support two injured astronauts, if necessary.

And they're people like Robert Paulin, a space flight technician at NASA's Glenn Research Center, who has worked on many of the scientific experiments flown on the shuttle. This includes wiring, soldering, assessing, and inspecting them to make sure that they meet NASA requirements and will work properly once in space.

Some technicians are involved in the design of space hardware. Computer-aided design operators, for example, create computer models of spacecraft components; the computers then drive automated machinery that actually build the components.

Communications technicians operate radio and other communications equipment, and test the equipment on the ground before it flies in space.

Radar technicians operate, test, and maintain radar equipment; laser technicians do the same for laser equipment, which is used to track satellites in orbit and measure from space the movement of Earth's crust, to name just two uses.

Optical technicians assist in the manufacturing and testing of optical components, such as camera and telescope lenses and mirrors and spacecraft windows.

Satellite control technologists monitor satellite operations, record and analyze data from satellites, and deal with problems that may arise on the satellite.

Quality assurance technologists examine hardware to make sure that it meets the necessary requirements for safety and effectiveness.

What to Do Now If You Want to Be a Technician

As you can see, the fields of expertise covered by technicians and technologists are enormous. How people end up in these careers varies enormously, too.

Glen Davis, for instance, joined the air force after high school, then taught welding for a while before being hired by Lockheed Space Operations Company. Attending classes in the evenings while working for Lockheed, he eventually earned a bachelor of science degree in professional aeronautics.

Robert Paulin admits that when he was in school, although he liked science, his favorite reading material was *Mad* magazine. But as an elementary school student, he always participated in science fairs and even won a couple of trophies. Then, in high school, he began taking electronics classes and, thanks to an excellent instructor, liked it so much it became his career, leading to his work today as a space flight technician.

Of all the opportunities to work in the space industry, the position of technician has the simplest requirements: basically, a high school diploma. However, the more training or experience you have in a field related to the job you'd like to have, the better your chance of getting that job. With only a high school diploma and no other training, you won't be doing any of the things the technicians described earlier in this chapter get to do—at least, not right away.

But one of the great things about working as a technician is that much of the training takes place on the job. Or, you can begin work as a technician and, like Glen Davis, go to night school to improve your education, opening up more advanced possibilities for you within the space industry.

The technicians I've mentioned in this chapter all have similar advice when it comes to pursuing a career as a space technician. "Always work to your highest ability in everything you do," says Barry Slack. "This will help you get the career you want."

"Study hard in school and try to be as diverse as possible in life," advises Ron Woods. "All of the experience that you gain will be used at some point in your life!"

Robert Paulin says, "Stay in school and keep those math and science grades up. Also read as much as you can on new technologies coming out. It changes so fast. It can be hard to keep up with it. The Internet is a great place to find out information. Use it."

Steven Van Meter maintains and operates several small robots at the Kennedy Space Center used to do work considered too dangerous for humans. Van Meter got his training on the job as a special agent for the department of energy, specializing in bomb disposal. His advice? "Education is the key. Mathematics is the universal language of the sciences and engineering. Because of that I would advise students to learn all that they can about mathematics."

In other words, you may be able to get a job as a technician with only a high school degree, but to really build a career as a technician you have to study, focusing on math and science. That's the most important thing you can do now if you'd like to work as a space technician someday.

And if being a technician doesn't sound very exciting, well, that's not how Glen Davis sees it. The nickname for technicians like him is "ground astronaut." He may not be traveling into space, but he's doing the next best thing.

Keeping the Public Informed: Communicators

You may think, after everything you've read so far, that all the careers in the space industry involve working with complex technology, that the sinews of space exploration are wires and cables and cold mechanical contraptions.

But while those things are obviously important to a high-tech endeavor like the space program, the true lifeblood of NASA and the many private companies involved in space exploration and exploitation is communication. Many of the people who work in the space industry are highly trained specialists in a particular technical or scientific field, as you've already seen. Maybe you have a particular bent in that direction yourself, and so those are the kinds of careers that interest you.

But maybe you're more like me. I've always been interested in science and technology, and when I was a kid I thought I'd be a scientist when I grew up; but along the way, I also became interested in writing. Other kids got their best marks on multiple-choice tests; I always got mine on essay tests. I read voraciously, both science and science fiction (among many other things), and by the time I was ten years old I was writing short stories.

The end result was that when I got out of high school and into college, instead of becoming a scientist or engineer, I became a writer, majoring in journalism and, once I graduated, working as a newspaper reporter for several years.

But I never lost my interest in science, and eventually I got a new job, as communications officer for the Saskatchewan Science Centre in Regina, Saskatchewan, where I had the opportunity to finally write extensively about science. I've been writing about science ever since.

The skills of the many writers and public relations people who work more directly in the space industry may be those that you have, too. The most important is the ability to take a large amount of complicated technical material and boil it down into a few well-chosen words that can be more easily understood by people who are not technical specialists themselves.

Vincent Mulhern is one of the writers who work full-time in the space industry; he's a technical writer with the Shuttle Safety and Mission Assurance Branch at NASA's Johnson Space Center in Houston.

Mulhern graduated in 1996 from the University of Houston with a degree in English. He'd always had an interest in space; his mother used to work at the Johnson Space Center as an engineer, training astronauts in how to deploy payloads from the shuttle. "I always thought she had the neatest job," he recalls. "When people asked what my parents did, I was proud to tell about it."

Despite that, Mulhern didn't expect to be working for NASA himself. "I didn't think there would be a place in the space program for someone with [an English] degree," he admits. "As graduation approached, however,

I began to learn about a field called technical writing. I had enjoyed math and science, though they were not my major, and I thought that I could understand and write about technical subjects."

He heard about an opening for technical writers at the Johnson Space Center, applied, and, thanks to his writing skill and grasp of technical subjects, got the job. "Working for a NASA contractor as a technical writer is a lot of fun," he says. "I'm thrilled to work with some great people and to be part of a team that achieves very exciting things. When I look out my office window and see rockets that first took Americans into space and to the Moon, or pass people like [astronaut] Shannon Lucid in the hall, I think, 'What could be a better place to work than this?'"

Lisa Doyle is another technical writer at NASA. She works at the Ames Research Center. She started out wanting to be a pilot, then went into computer science, then into aviation electronics, got a job as an electronics book club editor at a book publisher, and used that as a stepping stone to technical writing. "I always wanted to work for NASA," she says.

Although she is currently writing manuals on how to operate and fix wind tunnels at Ames Research Center, she notes, "My field is diverse, which means I can use my skills in several different technical fields and levels. It's dangerous to get too specialized in a field because that specialty could be phased out over time and all of your experience will be with that skill set." In short, she says, "Always continue to learn and grow."

The kind of communication that's important to the space industry doesn't just include writing manuals for people within the industry, which is the kind of technical writing Mulhern and Doyle do. Communication with people outside the industry is just as important, particularly in the case of NASA, which, as a government agency, has both a need and a duty to keep the public informed of its activities.

One of the people whose space career is focused on that aspect of communication is Lisa Malone, chief of the media services branch at NASA's John F. Kennedy Space Center since 1995. She manages a staff of sixteen people who provide information to news media all over the world about NASA's many different programs.

Malone has a lot of responsibilities in her position. For instance, she manages media activities surrounding launches, landings, and astronaut activities, and oversees all printed information, video releases, news conferences, official still photography, and live coverage of events. She works with movie and TV crews shooting footage to create NASA-related feature films. (When the film *Contact* was being shot, it was Malone who gave Jodie Foster a tour of the space center.) You may even have heard Malone's voice or seen her on television; she serves as a launch commentator for space shuttle and other rocket launches.

Essentially, whenever anyone in the public or the media needs information about what's going on at the John F. Kennedy Space Center, they contact Malone's staff. Providing that information is an enormous task. "The demand for information, from kids working on reports to TV news crews, far exceeds what we can provide," Malone says.

Malone worked her way up to her present position. She edited the in-house newsletter *Spaceport News* for three years; then for five years she was the lead space shuttle public affairs representative, and then news chief in the media service branch, where her jobs included assigning and editing news releases, fact sheets, press kits, public affairs plans, and status reports.

How did she launch her space career? Malone's college degree, like that of many other people who work in public relations positions within the space industry, is in journalism. She caught the "space bug" while she was attending the University of Alabama in Tuscaloosa. "I did an internship [at the Kennedy Space Center] in college and was dazzled by the cutting-edge technology and thrilling launches, so I stayed on full-time after graduation," she says.

Whether it's writing technical manuals used by researchers at a NASA or private lab, editing the internal newsletter for a space facility, or overseeing communication with the news media, there are many opportunities within the space industry for people who can communicate clearly.

What to Do If You Want to Be a Space Communicator

Read. Read, read, read. Then read some more. There's no better way to learn to use language well than to read.

Anything you read will help improve your ability both to understand what you read and to communicate it clearly to others, but if you're interested in working in the space industry as a communicator, then at least some of what you read should be related to space. Read science and science fiction books, read magazine articles about space, and read Web sites that focus on space and science issues. The more general knowledge you have about scientific and technical matters, and especially about what's happening in space, the better prepared you'll be for your eventual career as a space communicator. Take some formal classes in writing, attend workshops, or join a creative writing club, if one's available. Volunteer for the school newspaper. Nothing will tell you more quickly if you're communicating your ideas clearly than hearing what other people have to say about something you wrote!

Work hard in English, of course, but also keep up your grades in math and science. It wouldn't be necessary if your goal were to become a poet, but you want to be able to communicate technical information, and that means you have to be able to understand it well enough to explain it to someone else.

In college, journalism or English are the most likely subjects to major in. Public relations is another possibility. You may choose to study a particular technical subject, but it's not necessary. You may even choose not to study journalism, English, or public relations and still end up as a space communicator.

Doug Peterson, public affairs officer at NASA's Johnson Space Center, says, "In order to be successful in government, business, and NASA, you don't have to be a scientist, engineer, or technical person. There are a lot of other ways to get involved, other ways that people must be involved. I feel very strongly about the importance of liberal arts degrees. Liberal arts degrees help people understand and appreciate the full spectrum of ideas and issues that are related to all things."

Whatever education path you choose, you'll next want to look for job openings for technical writers or public affairs people with NASA or other companies working in the space industry. Remember Lisa Doyle's advice to be flexible and always continue to learn and grow, and you'll soon find that the demand for someone who can write clearly about technical subjects, whether in-house or for the general public, will lead you in new and exciting directions.

Two Routes to the Stars: Public and Private

8

Probably when you think of a space-related career, you think first of NASA. That's not surprising because NASA has been at the forefront of the space industry in the United States since it was created, and its manned space program, especially the Apollo lunar program in the 1960s, has captured the imagination of Americans and non-Americans alike for decades.

But increasingly, one can have a space-related career and never work for NASA. There are more and more private companies with an interest in space and more job openings for people who share that interest.

NASA in a Nutshell

In the late 1940s, the United States began pursuing research in rocketry and upper atmosphere science. President Dwight D. Eisenhower approved a plan to launch a scientific satellite sometime between July 1, 1957, and December 31, 1958, as part of the International Geophysical Year.

However, the Soviet Union beat the United States to the punch, launching *Sputnik I* on October 4, 1957. Shocked, the United States responded with greatly increased funding for scientific and technological programs, especially those related to aerospace engineering, and on January 1, 1958, the first American satellite, *Explorer 1,* was launched.

At the time, NASA didn't exist. However, there was an organization called the National Advisory Committee for Aeronautics (NACA), created in 1915 to "supervise and direct the scientific study of the problems of flight, with a view to their practical solutions."

One of the government's responses to *Sputnik I's* launch was to turn NACA into NASA to lead "the expansion of human knowledge of phenomena in the atmosphere and space." NASA came into existence on October 1, 1958.

Today, NASA employs close to 18,000 civil servants and has an annual budget of $14 billion. (That may sound like a lot, but in fact NASA's budget is less than 1 percent of the total federal budget and is the smallest budget of any of the federal government's major agencies.)

NASA's headquarters are in Washington, D.C. NASA's headquarters staff oversees NASA's four "strategic enterprises," which are space science (the scientific study of the universe), Earth science (the scientific study of Earth from space), human exploration and development of space (which oversees the space shuttle and International Space Station programs), and aeronautics and space transportation technology (which focuses on developing new technology for air and space transportation).

NASA has several affiliated facilities. The Ames Research Center in Moffett Field, California, develops aerospace technologies; conducts research in Earth, life, and space sciences; and develops information systems and technology for all NASA missions.

The Dryden Flight Research Center in Edwards, California, researches and develops advanced aeronautics and space technologies. Edwards Air Force Base is where many of NASA's experimental aircraft have been tested, and it is where the space shuttle lands when weather prevents it from landing in Florida.

The Glenn Research Center in Cleveland, Ohio, focuses on propulsion, electrical, and communication technology for aircraft and spacecraft.

The Goddard Space Flight Center in Greenbelt, Maryland, is home to a wide range of research on topics ranging from Earth science and astrophysics to satellite tracking and control.

The Jet Propulsion Laboratory in Pasadena, California, is operated by the California Institute of Technology under a NASA contract. Its focus is on planetary science. JPL is where many robotic spacecraft have been designed before being launched deep into the solar system, and it operates those spacecraft once they have been launched. JPL is also home of the worldwide Deep Space Network of large ground-based spacecraft communications dishes.

The Johnson Space Center in Houston, Texas, is world famous because it is the center for activities related to United States human space flight. This is where the space shuttle and International Space Station programs are managed, and where astronauts do much of their training.

The Kennedy Space Center in Florida is NASA's other very famous facility; it's where most launches take place of both space shuttles and nonreusable vehicles, and where the space shuttle usually lands.

The Langley Research Center in Hampton, Virginia, was established in 1917 as the nation's first aeronautical research laboratory, and it continues to conduct research and development of many different types of aerospace technology, relating to everything from airplane safety to satellites and space communication.

The Marshall Space Flight Center in Huntsville, Alabama, manages the space shuttle's propulsion systems—the rocket engines, external fuel tank, and solid-fuel boosters—and conducts research into a variety of space-related fields, from microgravity to biotechnology.

The Stennis Space Center in Massachusetts is the largest rocket propulsion testing facility in the United States, and the Wallops Flight Facility in Wallops Island, Virginia, launches NASA's smaller rockets and payloads, especially those that take only suborbital flights, rather than proceeding into orbit.

Applying for a Job with NASA

If you choose to pursue any of the career possibilities outlined in this book that involve working for NASA, you'll eventually have to apply for a job with NASA.

Fortunately, NASA has a Web site devoted exclusively to that topic at http://www.nasajobs.nasa.gov, telling you exactly what you have to do to apply for the many different types of jobs available with the space agency.

If you end up working directly for NASA (as opposed to working for one of NASA's contractors), you'll be a government employee, or civil servant.

To apply, you first find a job opening for which you think you're qualified. On the Web site mentioned above,

NASA lists both jobs that are open to the public and jobs that are open only to current NASA employees. Next, you prepare your résumé or fill out an application form, following the instructions given with the job announcement. Finally, you submit your résumé or application form to the office listed in the job announcement. And then you wait.

Applying for a Federal Job

Whenever you apply for a federal job, your résumé or application must contain certain general information in addition to the specific information requested in the job announcement. Here's what you have to include when applying for the first time.

Job Information

- Announcement number, title, and grade(s) of the job for which you're applying

Personal Information

- Full name, mailing address, and day and evening phone numbers (with area codes)
- Social security number
- Country of citizenship (most federal jobs require United States citizenship)
- Veterans' preference (military veterans receive special considersation over civilian applicants)

Education

- High school

Name, city, and state (zip code if known)
Date of diploma or GED
- Colleges and universities
Name, city, and state (zip code if known)
Majors
Type and year of any degree received (if you didn't earn a degree, provide the total number of credits you earned and indicate whether they're based on semester or quarter hours)
A copy of your college transcript if (and only if) the job vacancy announcement requests it

Work Experience

You have to provide the following information for your paid and nonpaid work experience related to the job for which you are applying:
- Job title
- Duties and accomplishments
- Employer's name and address
- Supervisor's name and phone number
- Starting and ending dates (month and year)
- Hours per week
- Salary
- Indicate if the government may contact your current supervisor

Other Qualifications

- Job-related training courses (title and year)
- Job-related skills; for example, other languages, computer software/hardware, tools, machinery, typing speed

- Job-related certificates and licenses (current)
- Job-related honors, awards, and special accomplishments; for example, publications, memberships in professional or honor societies, leadership activities, public speaking, and performance awards (give dates but do not send documents unless requested)

As a federal employee, you'll receive a good salary and benefits. And NASA has a very low turnover rate, which reflects a high degree of employee satisfaction and the dedication of NASA employees to the agency's exciting goals of exploring and exploiting space. After all, if you want a career in space, where can you turn besides NASA?

As it happens, you can turn to the private sector, including both contractors who work directly with NASA to run its programs, and companies that are involved in space but only indirectly, or not at all, with NASA.

The Private Space Sector

Because NASA has the only manned space program in the country, and manned space flight—and therefore NASA—gets all the headlines, it's easy to forget that NASA doesn't operate in a vacuum. It depends on private companies to support many of its programs.

The space shuttle is actually operated and maintained by a private company. United Space Alliance, a joint venture of Lockheed Martin and Boeing, took over operation of the space shuttle fleet in 1996, and today provides mission operations and planning, training, software development, logistics, launch operations, and more.

Its two parent companies also remain heavily involved in the space program. Boeing, the company that makes airplanes like the Boeing 747 jumbo jet, is the prime U.S. contractor for the International Space Station. It also supplies launch services, satellites, and propulsion systems. Lockheed Martin builds both government and commercial satellites and launch vehicles.

There are hundreds of smaller companies that NASA contracts to provide equipment and services for all of its space activities, from the shuttle to scientific satellites to interplanetary probes—all of them possible places for you to find a career in a space-related field.

But there are also hundreds of companies that are involved in the space industry without necessarily being involved with NASA's programs or the space programs of other countries.

Most of these companies work in the satellite industry, and most of the satellites they work on are communications satellites. Broadcast satellites, satellites for transmitting digital audio radio, and even satellites for providing Internet connections are among the fastest growing sectors of the space industry. That translates into lots of jobs in those sectors for people with the necessary technical training.

How big is the private space industry? Worldwide revenues were estimated at $87 billion in 1999 and $97 billion in 2000. Worldwide, 1.1 million people were employed in the private space sector in 1999.

The International Space Business Council projects that by 2005, worldwide revenues will almost double to $168 billion; by 2010, revenues could be between $200 billion and $300 billion.

Careers in Outer Space: New Business Opportunities

One reason is that increasingly all types of communication involve satellites. As Scott Sacknoff, president of the International Space Business Council, says, "If, in the course of the week, you get gas and pay by credit card, get cash from an ATM, fill a prescription at a national drugstore chain, or buy or sell stock, chances are there may be a satellite involved."

There's another area of private space enterprise that is also growing that's not communications-related. Called remote sensing, it's the use of satellites to capture data about Earth—primarily images, but not necessarily those created with available light. Satellites also photograph Earth using radar, infrared light, and other frequencies of electromagnetic radiation, all of which can provide important information about geological formations, the health of forests, crop coverage, and more.

If you're interested in a career within the private space industry, you need the same kind of engineering or technical training you'd need to land a technical job within NASA. Study math and science, keep your grades up, read about space, and choose a directly related field of study at university. Then find a company that's doing the sorts of things in space that interest you and apply.

With billions of dollars of revenue and more than a million employees now worldwide, and extensive growth expected over the next few years, the private space industry offers many more opportunities for space-related careers than the fairly limited civil-service ranks of NASA. Outside of the United Space Alliance, though, not a lot of it is directly associated with human space flight.

However, as we'll see in the next chapter, that may be about to change.

The Future in Space

9

Dennis Tito was a teenager in Queens, New York, when the Soviet Union launched the first artificial satellite, *Sputnik I,* on November 4, 1957. The launch sparked his imagination and filled him with a desire to one day travel in space.

Although he didn't become an astronaut, he did the next best thing, pursuing a career in space by earning bachelor's and master's degrees in aerospace engineering. In 1964, he went to work at the Jet Propulsion Laboratory in Pasadena, California, where, among other things, he charted the flight paths for NASA's first probes to Mars, the Mariners.

But although his $15,000-a-year salary was pretty good for the 1960s, Tito set out to make more money than he could at NASA, founding his own business, Wilshire Associates, in the 1970s. He used his mathematical ability to analyze the stock market instead of the trajectories of spacecraft, and by the time he turned forty, he was a millionaire. Today he's a multimillionaire, with a personal fortune estimated at $200 million.

But even though he'd given up a career in the space program for a career in finance, Tito hadn't given up his dream of someday flying in space. That dream began to seem like a possibility in the early 1990s. In 1990, the Japanese television network TBS paid the Soviet Union $28 million to send journalist Tohiro Akiyama to the *Mir* space station for a week. In 1991, a British consortium arranged a similar deal for British chemist Helen Sharman, who was chosen in a competition entered by 13,000 people.

Tito seriously considered approaching the Soviet Union to fly him to *Mir*, but then the Soviet Union collapsed and the opportunity along with it.

Then, in April 2000, a company called MirCorp, formed to find commercial uses for the *Mir* space station, called Tito. Would he like to fly to *Mir*?

He jumped at the chance. He put millions of dollars into an account that the Russian space program would be able to access once he was launched to *Mir*, and he moved to Star City, the cosmonaut headquarters outside Moscow.

But *Mir* was aging, suffering more and more problems. The Russians decided to de-orbit it before Tito had an opportunity to fly to it. They made him an alternate offer: a trip to the still-under-construction International Space Station. Russian Soyuz spacecraft serve as the station's lifeboats, and they have to be replaced every six months. Two cosmonauts are required to pilot the craft to the station, but each Soyuz spacecraft has three seats. If Tito wanted, he could ride in the third seat.

Tito agreed, and despite opposition from NASA and other partners in the International Space Station, who didn't like the idea of an "amateur" on board the station, something they saw as a potential hazard, Tito launched into orbit on April 28, 2001. He spent six days on board the International Space Station and returned safely to Earth.

"The personal experience went well beyond my dreams," Tito said on his return; in fact, he thought it was "ten times better" than he expected. But, he added, he doesn't want to make the trip again: "I want other people to make it instead." There are a lot of people who would like to. And there are a lot of other people who are looking at ways to make it possible.

Tourists in Space Mean Careers in Space

If and when space tourism takes off, there'll be a whole new range of space careers to explore, from space-liner pilot to zero-gravity gourmet cook.

Zero-gravity gourmet cook? Absolutely. After all, visitors who have spent thousands or even hundreds of thousands of dollars to spend a week at a space hotel, or ride a space liner on a sightseeing tour around the Moon, aren't going to want to eat ordinary food.

And if space hotels and sightseeing tours around the Moon sound like something that will never happen, someone should tell the companies that are spending serious money discussing those very possibilities.

Hilton Hotels Inc. announced in 1999 that it is looking into the feasibility of a space hotel. "We want to take a hard look at it and see if Hilton can be the first in space," was the way Hilton spokeswoman Jeannie Datz put it.

"Is this for young, really healthy people or can the John Glenns of the world go up there and have a good experience?" Datz asked. And—to bring us back to that zero-gravity gourmet cook career you may have your eye on—"what are you going to eat? If you want your New

York steak or pasta primavera, is it going to be available, or is it going to be in pill or freeze-dried form?"

Another company looking at building a space hotel is the Space Island Group of West Covina, California, which has suggested using empty space-shuttle fuel tanks for that purpose. Guests could take space walks, gaze down at Earth (already an endlessly popular pastime for astronauts), or try their hands at space gardening.

More than just hotels are being talked about. Robert Bigelow, owner of the Las Vegas–based Budget Suites of America lodging chain, has put $500 million of his money into founding a new company, Bigelow Aerospace, which has the long-term goal of building a cruise ship that would fly tourists from Earth to the Moon and back. This would open up whole new career opportunities in fields like "space flight cabin attendant" and "space liner entertainment director." Or even, "zero-gravity aerobics instructor."

If all that sounds a little farfetched, there are less technologically challenging forms of space tourism that could be up and running within a couple of years.

The X Prize

An organization called the X Prize Foundation is offering a $10 million prize to the creators of the first privately funded spacecraft capable of carrying three humans to an altitude of at least 100 kilometers on two consecutive flights within two weeks.

The idea is to stimulate the creation of a new generation of launch vehicles designed to carry passengers into space cheaply, quickly, and safely, kick-starting the space tourism industry. The X Prize is modeled after the early aviation prizes that did so much to spur the development of airplanes; in fact, the X Prize Foundation is located in St. Louis partly because of the city's connection with Charles Lindbergh, who flew solo across the Atlantic Ocean in the *Spirit of St. Louis* in 1927 to win the $25,000 Orteig Prize.

The private-funding requirement ensures that a government can't come in and win the competition by throwing money at it. The 100-kilometer altitude was chosen because it's above the 80-kilometer altitude that the U.S. Air Force recognizes as worthy of astronaut wings, but not so high that the reentry speed requires exotic heat-shielding materials. The three-person passenger capability means the winning spacecraft can be immediately turned into a profit-generating tourist vehicle. And the two-week reflight requirement is designed to keep costs down, because the winning spacecraft cannot be one that must be rebuilt after each flight. Only refueling and worn parts replacement are allowed.

As this book was being written, there were twenty teams competing for the X Prize, from the United States, Canada, Russia, Argentina, and the United Kingdom.

One of the front-runners, Canada's daVinci Project, demonstrates the innovative thinking the teams are bringing to the goal of more accessible human space flight. The daVinci Project spacecraft will start its journey at the end of a 1,000-foot tether attached to a 25-story hot-air balloon. At 40,000 feet, the liquid oxygen/kerosene engine of the 24-foot spacecraft will fire, first propelling it at an angle, to ensure that it clears the balloon, then swiveling to send it straight up to well over the 100-kilometer minimum altitude. Maximum speed will be four times the speed of sound.

As the spacecraft descends, a "ballute"—a large inflatable cone, a cross between a balloon and a parachute—will deploy over the rocket's descending end, stabilizing it and protecting it from the heat of reentry. At a lower altitude, a flyable parafoil will deploy, and the rocket will glide down to a landing zone somewhere in western Canada.

By the time you read this, the daVinci Project, or some other entry in the X Prize competition, may have already succeeded in placing the first private astronaut in space, and it may be very close to taking paying customers up for short suborbital jaunts, as well.

Career opportunities? There's everything from engineers and technicians, just as with NASA's manned space flight program, to pilot, to all the support jobs that go with any tourist-based enterprise: public relations, promotion, graphic design, and so on.

Other Commercial Possibilities

Space tourism gets a lot of attention from people interested in seeing more humans in space because the tourism industry seems like one that could generate the money needed to make it happen. But once you have cheaper and easier ways to get humans into space, many more commercial opportunities open up.

A report by Kelly Space & Technology, Inc., on the demand for space transportation between 2010 and 2030, for instance, identified many possibilities, including:

Space Burials

Imagine having your ashes shot into orbit, to remain a permanent satellite of Earth. Career opportunity: space funeral director.

In-Orbit Manufacturing

There have already been a number of tests of in-orbit manufacturing. The lack of gravity could make it easier to manufacture exotic new materials or drugs. Career opportunities range from the scientific and technical to being the astronaut that goes up to service the manufacturing satellites.

In-Orbit Entertainment

Imagine film studios on orbiting space stations, or even zero-gravity theme parks. Imagine working as a film director on a space station or wearing Mickey Mouse ears on your space suit!

Space Sports

What kind of sports could be developed to take advantage of the lack of gravity and complete freedom of movement? Space polo? Space football? Or something brand-new? Career opportunities: pro space athlete, space-sports referee, space-sports play-by-play or color commentator.

Space Medical Facilities

With enough people in orbit, an orbiting hospital might be needed. Why risk sending someone down to Earth? A space hospital would need trained medical personnel with a whole new set of skills, so they could work effectively in zero gravity.

Space Utilities

Imagine giant satellites circling Earth, soaking up the energy of the Sun and beaming it down to Earth to power factories, offices and homes. Career opportunities on both the ground and in space would open up; somebody's got to maintain both the satellites and the ground stations.

Space Couriers

Right now, if you need to get a package halfway around the world within a couple of hours, you're out of luck. A suborbital hypersonic spaceplane could do it. Various companies are trying to develop one. You could end up working for them.

Space Rescue

Wherever people go, accidents happen. Once there are lots of civilians in space, personnel will have to be trained and standing by to perform space rescues. Maybe one of those people will be you.

Space Salvage

There's already a lot of junk orbiting Earth at very high speeds—everything from bits of old satellites to tools dropped by astronauts. Eventually, we may have to try to retrieve some of that, both for safety reasons and because some of it could be valuable as collectors' items, if nothing else. ("Own a piece of a genuine Soviet spy satellite from 1974!") It could be hard, dangerous work—as salvage work always is—but it could be your ticket into space.

How many people could be going into space in the next ten or twenty years? Kelly Space & Technology projects as many as 10,000 could take suborbital flights from 2010 to 2020, and as many as 4,000 could spend some time in orbit from 2015 to 2025—and that's just tourists.

Careers in Deep Space

But what about moving beyond Earth's orbit?

There are also individuals and organizations that would like to see tourism and other industries flourishing not just in orbit, but on the Moon and, eventually, even on Mars.

The Moon is attractive because it's relatively close, and we've been there before, though not since 1972, when the astronauts of *Apollo 17* visited.

A permanent base on the Moon could be an even more attractive tourist attraction than orbit. The view of Earth wouldn't be as good, but there are more tourist attractions to visit—the *Apollo 11* landing site, for instance. And the Moon's gravity, one-sixth that of Earth's, might even be a selling point. There would not be as much chance of space sickness, but it could provide lots of fun of the see-how-high-I-can-jump and can-you-believe-how-far-I-hit-that-golf-ball? variety. (Golf on the Moon? Why not? Alan B. Shepard Jr., commander of the *Apollo 14* mission, used a makeshift golf club to hit a golf ball on the Moon way back in 1971.) Career opportunities: everything from lunar tour guide to Moon caddy.

Whether any of these predictions come true, the point is that we will continue to move into space, both ourselves and our machines. And that means that more and more careers, many of which we can't even imagine right now, will be opening up, both in space and here on Earth.

If you want to be part of the space industry, if you want to help humanity make the move into space, and especially if you someday dream of traveling to space yourself, then you're living at the right time.

Humanity's future is in space. Maybe yours is, too.

Timeline: Important Dates in Space Exploration

March 16, 1926 Robert Goddard launches the world's first successful liquid-fueled rocket, in Auburn, Massachusetts.

October 3, 1942 The first German V-2 rocket is successfully launched.

1946 The United States and the Soviet Union begin their space research programs by experimenting with captured German V-2s with the help of captured German rocket scientists.

April, 1955 The Soviet Union announces plans to explore the Moon and devise an Earth-orbiting satellite.

July 29, 1955 President Dwight D. Eisenhower announces plans to launch U.S. satellites.

October 4, 1957 The Soviets launch the first artificial satellite, *Sputnik 1*.

November 4, 1957 The Soviets launch *Sputnik 2*, carrying a dog, Laika, the first animal to orbit Earth.

January 31, 1958 The United States launches its first satellite, *Explorer 1.*

October 1, 1958 The National Aeronautics and Space Administration (NASA) begins operation.

January 2, 1959 The Soviets launch the *Luna 1* space probe, the first human-made object to fly past the Moon. A follow-up probe crash-lands on the Moon; another sends back pictures of the Moon's far side.

April 9, 1959 The United States introduces its first seven astronauts.

April 12, 1961 Soviet Yuri Gagarin is the first person in space, orbiting Earth in a 108-minute flight.

May 5, 1961 Astronaut Alan Shepard is the first American in space, completing a fifteen-minute suborbital flight.

February 20, 1962 John Glenn is the first American to orbit Earth.

June 16, 1963 Soviet Valentina Tereshkova is the first woman in space, orbiting Earth in *Vostok 6.*

March 18, 1965 Soviet Alexei Leonov is the first person to walk in space.

June 3, 1965 Ed White makes the first U.S. space walk.

July 14, 1965 The U.S. probe *Mariner 4* transmits the first close-range images of Mars back to Earth.

February 3, 1966 The Soviet *Luna 9* probe makes the first soft landing on the Moon.

June 2, 1966 The U.S. *Surveyor 1* probe lands on the Moon.

January 27, 1967 Astronauts Virgil "Gus" Grissom, Ed White, and Roger Chaffee are killed during a test of *Apollo 1.*

April 24, 1967 Soviet Vladimir Komarov dies when *Soyuz 1* crashes during reentry.

December 21, 1968 The *Apollo 8* crew, Frank Borman, Jim Lovell, and Bill Anders, are the first humans to orbit the Moon.

July 20, 1969 Astronaut Neil Armstrong of *Apollo 11* becomes the first person to walk on the Moon.

April 11, 1970 The crew of *Apollo 13* narrowly escapes death after an explosion.

April 19, 1971 The *Salyut 1* space station is launched by the Soviets. Its first team of cosmonauts dies during reentry in June.

December 2, 1971 The Soviet Union's *Mars 3* makes the first soft landing on Mars.

December 7, 1972 *Apollo 17* becomes the last mission to visit the Moon to date.

May 14, 1973 Skylab, America's first space station, is launched. It remains in orbit for six years.

July 17, 1975 *Apollo 18* and *Soyuz 19* dock in space, the first international collaboration.

July 20, 1976 The U.S. *Viking 1* orbits Mars and lands a craft on its surface that conducts soil analysis. *Viking 2* repeats the feat in September.

February 18, 1977 NASA begins testing the space shuttle.

April 12, 1981 John Young and Robert Crippen fly the first space shuttle, *Columbia*, into orbit.

November 12, 1981 *Columbia*'s second flight marks the first time any spacecraft has returned to space.

June 13, 1983 The *Pioneer 10* space probe becomes the first man-made object to leave the solar system.

June 18, 1983 Sally Ride is the first U.S. woman in space.

August 30, 1983 Guion Bluford is the first African American astronaut to fly in space.

February 7, 1984 Bruce McCandless takes the first untethered space walk.

January 28, 1986 Space shuttle *Challenger* explodes after launch, killing its crew.

February 20, 1986 The Soviets launch the core of their *Mir* space station.

September 29, 1988 The first space shuttle since the *Challenger* disaster is launched.

April 25, 1990 The shuttle deploys the Hubble Space Telescope. Its mirror proves to be faulty.

December 4–10, 1993 Astronauts capture and repair the Hubble Space Telescope.

February 3, 1994 Cosmonaut Sergei Krikalev is the first Russian to be launched in a U.S. space shuttle.

March 14, 1995 Norman Thagard is the first American to be launched on a Russian rocket.

March 25, 1995 Cosmonaut Valery Polyakov sets a space-endurance record of 437 days, 18 hours aboard *Mir*.

September 7, 1996 American Shannon Lucid sets an endurance record for women in space with a 188-day mission aboard *Mir*.

June 25, 1997 The U.S. *Sojourner* Rover becomes the first vehicle to roam Mars.

October 29, 1998 John Glenn, seventy-seven, is the oldest person to visit space, thirty-six years after his first flight.

November 20, 1998 Russia launches *Zarya,* the first element of the International Space Station.

December 4, 1998 The first U.S. module of the International Space Station, Unity, is attached to *Zarya.*

November 2, 2000 The first permanent crew boards the International Space Station.

March 23, 2001 The fifteen-year-old Russian space station *Mir* returns to Earth in a controlled descent, splashing into the Pacific Ocean.

April 28, 2001 American Dennis Tito is the first space tourist, paying the Russian government $20 million for a six-day trip to the International Space Station.

Glossary

aeronautics Science dealing with the operation of aircraft.

aerospace Science that involves both the atmosphere and outer space. A company that builds airplanes and satellites, for instance, would be an aerospace company.

asteroid A small, rocky object that orbits the Sun, especially (though not exclusively) between the orbits of Mars and Jupiter.

astronaut American term for a person who travels to outer space.

astronomy The study of objects and matter outside Earth's atmosphere.

bachelor's degree The basic college degree, usually awarded after four years of study, based on a combination of general-knowledge courses and courses focusing on the particular discipline in which the degree is awarded.

biology The study of living organisms.

chemistry The study of the composition, structure, and properties of substances and the transformations that they undergo.

civil servant Someone employed by the government.

climatology The scientific study of climates.

comet A Sun-orbiting object that, as it approaches the Sun, begins to give off gases and dust, usually resulting in a long tail that points away from the Sun.

computer scientist A computer professional who designs computers and software, and develops ways to apply computers to new uses.

contractor A person or company that is contracted to provide services or supplies to another person or company. NASA uses contractors extensively to build and operate facilities and equipment.

cosmonaut The Russian word for someone who travels to outer space.

doctorate The highest degree granted by a university, usually after several years of additional research and study after college graduation.

EVA Extravehicular activity; term used by NASA for activities carried out by astronauts in space outside of their spacecraft; also known as a space walk.

geology The study of the history of Earth (or other planets), especially as recorded in rocks.

gravity A force generated by anything that has mass that attracts anything else that has mass. Gravity is what keeps Earth orbiting around the Sun, the Moon orbiting around Earth, and us from floating off into outer space.

logistics The handling of the details of an operation.

master's degree A postgraduate degree awarded by a university to someone who has done additional study and independent research in a particular field.

mathematics The science of numbers.

meteorology The study of the atmosphere and its phenomena, especially weather and weather forecasting.

microgravity The proper term for the lack of gravity astronauts experience while in orbit in space, which causes objects to float. NASA prefers this term to "weightlessness" or "zero gravity" because gravity is still active; it just isn't being felt.

mission specialist Astronaut responsible for coordinating onboard operations. Mission specialists help plan the crew's activities; monitor the use of the shuttle's food, water and fuel; and conduct experiments and EVAs.

NACA The National Advisory Committee on Aeronautics, created in 1915 to "supervise and direct the scientific study of the problems of flight, with a view to their practical solutions." NACA became NASA in 1958.

NASA The National Aeronautics and Space Administration, formed October 1, 1958, from NACA to further "the expansion of human knowledge of phenomena in the atmosphere and space."

nutritionist Someone who studies the processes by which an animal or plant takes in and utilizes food substances.

oceanography The study of the oceans.

orbit The path one object makes when it revolves around another, as space craft do around Earth.

pad rat NASA slang for the technicians who service spacecraft when they're on the launching pad, as well as the machinery of the launching pad itself.

payload Anything carried by an aircraft or spacecraft that is necessary to the purpose of the flight but isn't necessary to the operation of the craft, e.g., a satellite or scientific instrument.

payload specialist An astronaut specially trained to look after and operate the space shuttle's payload on a particular mission.

physics The study of matter and energy and their interactions.

pilot astronaut An astronaut who commands the space shuttle and is responsible for its overall safety and operation, any in-space maneuvering, and its landing.

psychiatrist A medical doctor who specializes in mental, emotional, and behavioral disorders.

remote sensing The use of instruments on satellites and aircraft to obtain data about Earth.

robot A machine that carries out tasks one might ordinarily expect a human to perform.

rocket The basic propulsion system for space flight, consisting of a combustion chamber in which fuel and oxygen are burned, and a nozzle through which the resulting exhaust is directed, propelling the rocket forward.

satellite Any object that orbits another object in space.

space adaptation syndrome The collection of symptoms displayed by humans who spend an extended period of time in space; for example, an increase in height, bloating, nausea, loss of bone density, etc.

spaceplane A vehicle that can take off like an airplane, travel into space, then return to Earth and land like an airplane.

space probe An unmanned spacecraft sent into space to gather scientific data about space phenomena.

space sickness Nausea brought on by microgravity conditions in space.

space station A permanent crewed outpost in space.

space tourist Someone who pays out of his or her own pocket to travel into space.

space walk Extravehicular activity; term used by NASA for activities carried out by astronauts in space outside of their spacecraft.

statistics A branch of mathematics dealing with the collection, analysis, interpretation, and presentation of masses of numerical data.

suborbital An adjective for anything that leaves Earth's atmosphere, but because of the speed and trajectory at which it is traveling, does not go into orbit around Earth, but instead falls back into the atmosphere.

systems analyst A computer scientist specializing in the way many different computers work together in a network.

technician Someone with specialized technical skills; for instance, a space suit technician is someone who knows a great deal about how to maintain and operate a space suit.

trajectory The path that an object follows in space.

V-2 An unmanned guided ballistic missile developed by Germany during the Second World War and used to attack Allied targets. Captured V-2s formed the basis of postwar missile and space research by both the United States and the Soviet Union.

For More Information

In the United States

NASA Headquarters
300 E. Street SW
Washington, DC 20546-0001
(202) 358-0000
e-mail: info-center@hq.nasa.gov
Web site: http://www.nasa.gov

In Canada

Canadian Space Agency
Communications Directorate
6767 route de l'Aéroport
Saint-Hubert, PQ J3Y 8Y9
(450) 926-4800
Web site: http://www.space.gc.ca

Web Sites

Institute for Research—Careers in Space and Aerospace
Science
http://www.careers-internet.org/aerospace1.htm

International Space Business Council
http://www.spacebusiness.com

Look to the Future: Careers in Space
http://mgs-mager.gsfc.nasa.gov/Kids/careers.html

NASA Jobs Homepage
http://www.nasajobs.nasa.gov

Occupational Outlook Handbook, 2000–01 Edition
http://stats.bls.gov/ocohome.htm

Space Careers
http://www.spacelinks.com/SpaceCareers

Space Jobs Inc.
http://www.spacejobs.com

Spacelink—Careers
http://spacelink.nasa.gov/Instructional.Materials/
   Curriculum.Support/Careers

X PRIZE Foundation
http://www.xprize.org

For Further Reading

Maze, Stephanie. *I Want to Be an Astronaut.* San Diego: Harcourt Brace, 1997.

Pasternak, Ceel, and Linda Thornburg. *Cool Careers for Girls in Air and Space.* Manassas Park, VA: Impact Publishers, 2001.

Sacknoff, Scott, and Leonard David. *The Space Publications Guide to Space Careers.* Bethesda, MD: Space Publications, 1998.

Sheffield, Charles, and Carol Rosin. S*pace Careers.* New York: Morrow, 1984.

Stine, G. Harry. *Living in Space: A Handbook for Work and Exploration Beyond the Earth's Atmosphere.* New York: M. Evans & Co., Inc., 1997.

Index

Index

About the Author

Edward Willett was born in New Mexico, grew up in Texas, and now lives in Regina, Saskatchewan, Canada, with his wife, Margaret Anne, and daughter, Alice. He worked as a newspaper reporter and editor and as communications officer for the Saskatchewan Science Centre before becoming a full-time freelance writer. He's the author of more than a dozen books, including young adult science fiction and fantasy novels, children's science books, and computer books; writes a weekly science column for newspapers and radio; and hosts a local TV program about computers. He's also a professional actor and singer. You can visit him online at www.edwardwillett.com.

Series Design
Danielle Goldblatt

Layout
Tahara Hasan